Cover and internal design by Will Riley

Published by Sourcebooks eXplore, an imprint of Sourcebooks Kids
P.O. Box 4410, Naperville, Illinois 60567-4410
(630) 961-3900
sourcebookskids.com
First published as Red Kangaroo's Thousands Physics Whys: *Sir Isaac Newton Knows*
in 2018 in China by China Children's Press and Publication Group.
Library of Congress Cataloging-in-Publication Data is on file with the publisher.
Source of Production: PrintPlus Limited, Shenzhen, Guangdong Province, China
Date of Production: February 2020
Run Number: 5017784
Printed and bound in China.
PP 10 9 8 7 6 5 4 3 2 1

Let's Get Moving!

Speeding into the Science of Motion with Newtonian Physics

sourcebooks eXplore

#1 Bestselling
Science Author for Kids
Chris Ferrie

Red Kangaroo loves to play with her ball. When she throws it up, it always comes down. Sometimes it moves fast, and sometimes it moves slowly. Red Kangaroo wonders why.

"Dr. Chris will be able to help me!" she says.

Red Kangaroo finds Dr. Chris in his lab. “Can you tell me what makes things move the way they do?” she asks him.

“Things that move follow three rules,” Dr. Chris replies. “A famous physicist named **Sir Isaac Newton** discovered these rules, so we call the rules Newton’s Laws of Motion.”

"I want to know these rules!" cries Red Kangaroo. "Can you please teach them to me, Dr. Chris?"

“Yes! But first I need to teach you three very important words that will help you understand Newton’s Laws. They are force, mass, and acceleration.

“Let’s start with force,” says Dr. Chris. “**Force** is any push or pull on something. An example of this is pushing a door open or pulling it shut.”

Acceleration

Dr. Chris

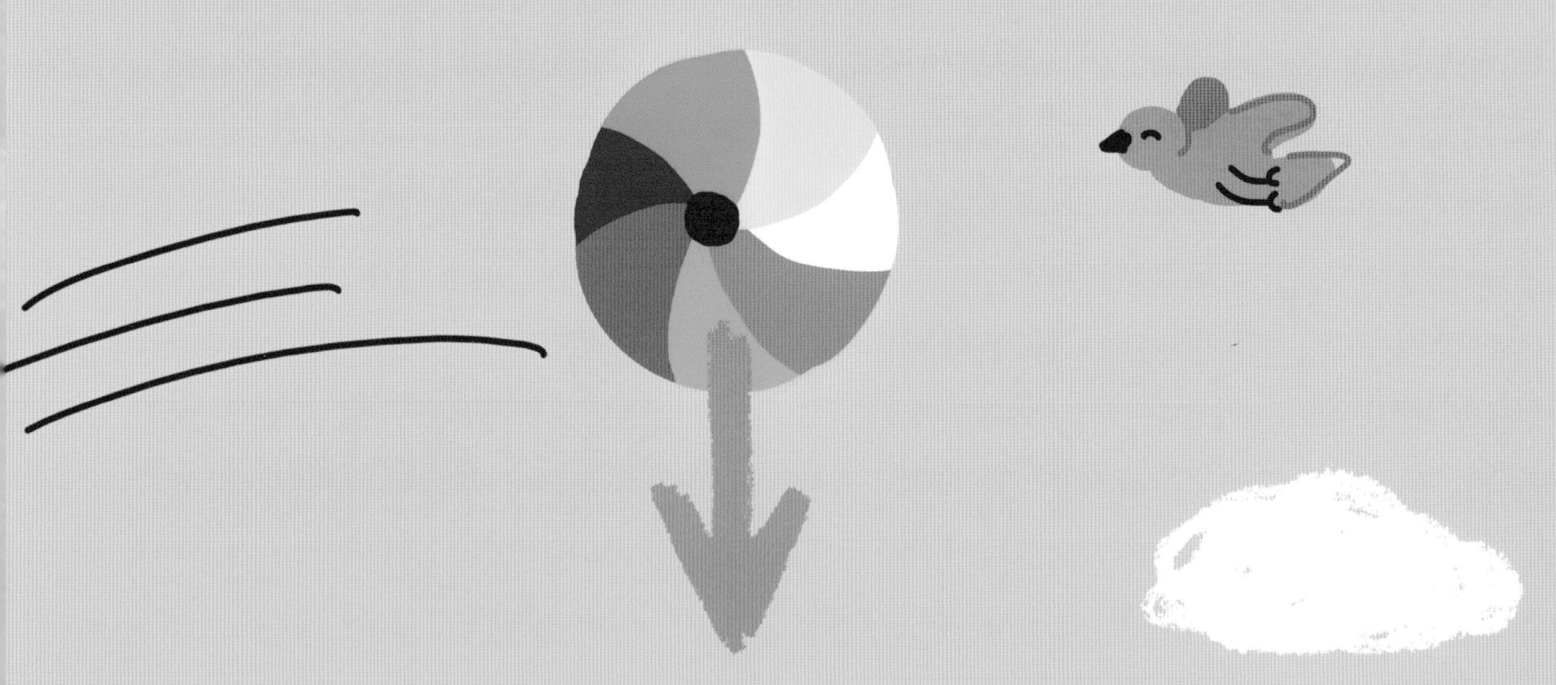

“So I am using force when I throw my ball into the air? My force pushes it up?” Red Kangaroo asks.

“Exactly! And **gravity** is the force that pulls it back down,” Dr. Chris responds. “Gravity is a very important pulling force. Without gravity, everything would stay up in the air.”

Dr. Chris

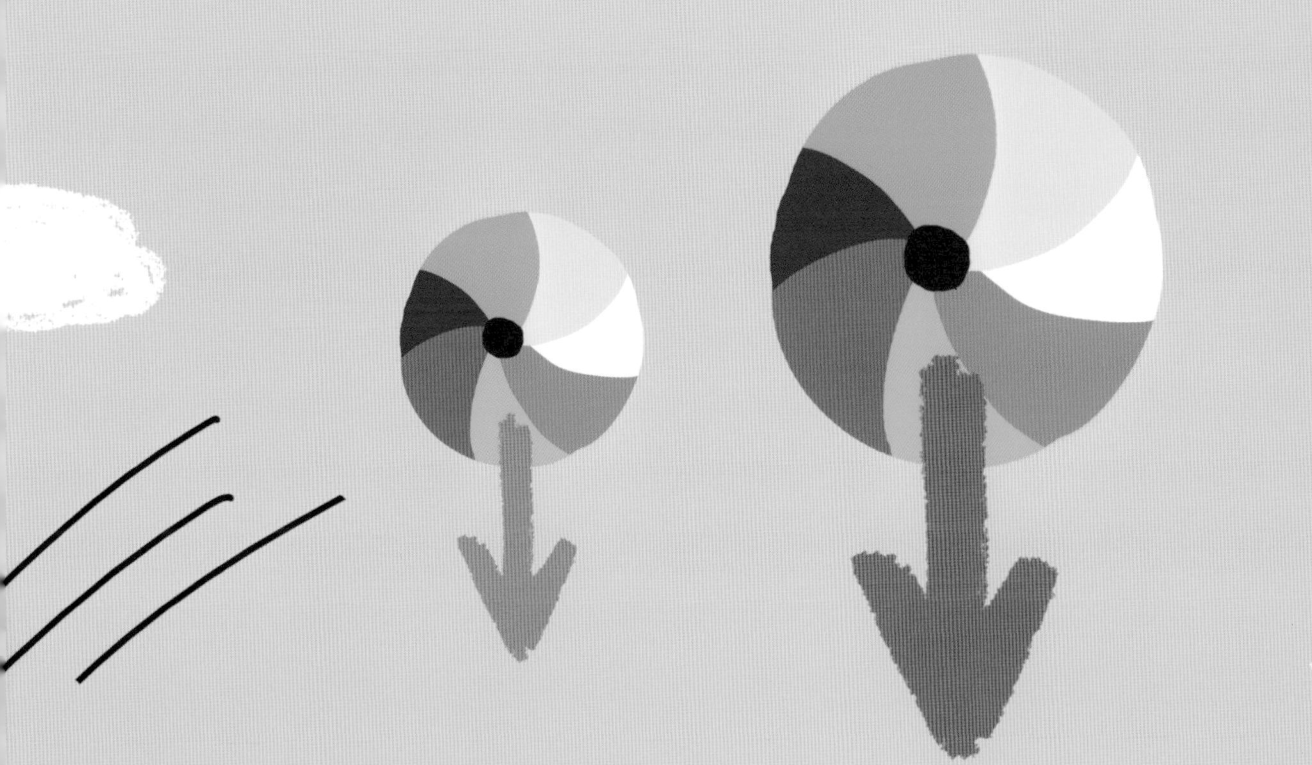

“Next, we have **mass**,” Dr. Chris continues. “Mass is how much matter something is made of.”

“I get it! The bigger ball has more mass than the smaller ball!” Red Kangaroo says.

“Yes! Very good!” says Dr. Chris. “And we can connect mass to force. Let’s throw both balls in the air. The ball with the greater mass will feel more force from gravity.”

“Finally, we have **acceleration**. This is change in speed,” says Dr. Chris. “A ball is accelerating if it starts out moving slowly and then goes faster.”

Dr. Chris

“I understand force, mass, and acceleration!” says Red Kangaroo. “Am I ready to learn Newton’s Laws now?”

“Yes, you are! Let’s go back to the lab,” replies Dr. Chris.

Dr. Chris

Dr. Chris

“**Newton’s First Law** says that an object will either stay still or continue moving at the same speed and in the same direction unless a force acts on it.”

“So this ball is not moving and will stay that way unless I push it?” asks Red Kangaroo.

“Yes! When you push the ball, you use your force to make it move,” says Dr. Chris. “The ball will continue to move at the same speed until my hand or another force, like the wall, stops it.”

Dr. Chris

Dr. Chris
2

“Now let’s go on to **Newton’s Second Law**. This law says that an object with a greater mass will need more force to accelerate it.”

"I pushed both of these balls with the same amount of force," says Red Kangaroo. "Does Newton's Second Law mean the smaller ball will accelerate faster because it has less mass?"

"You've got it, Red Kangaroo!" Dr. Chris says. "But you can also have them accelerate at the same speed. You would just need to use more force on the bigger ball!"

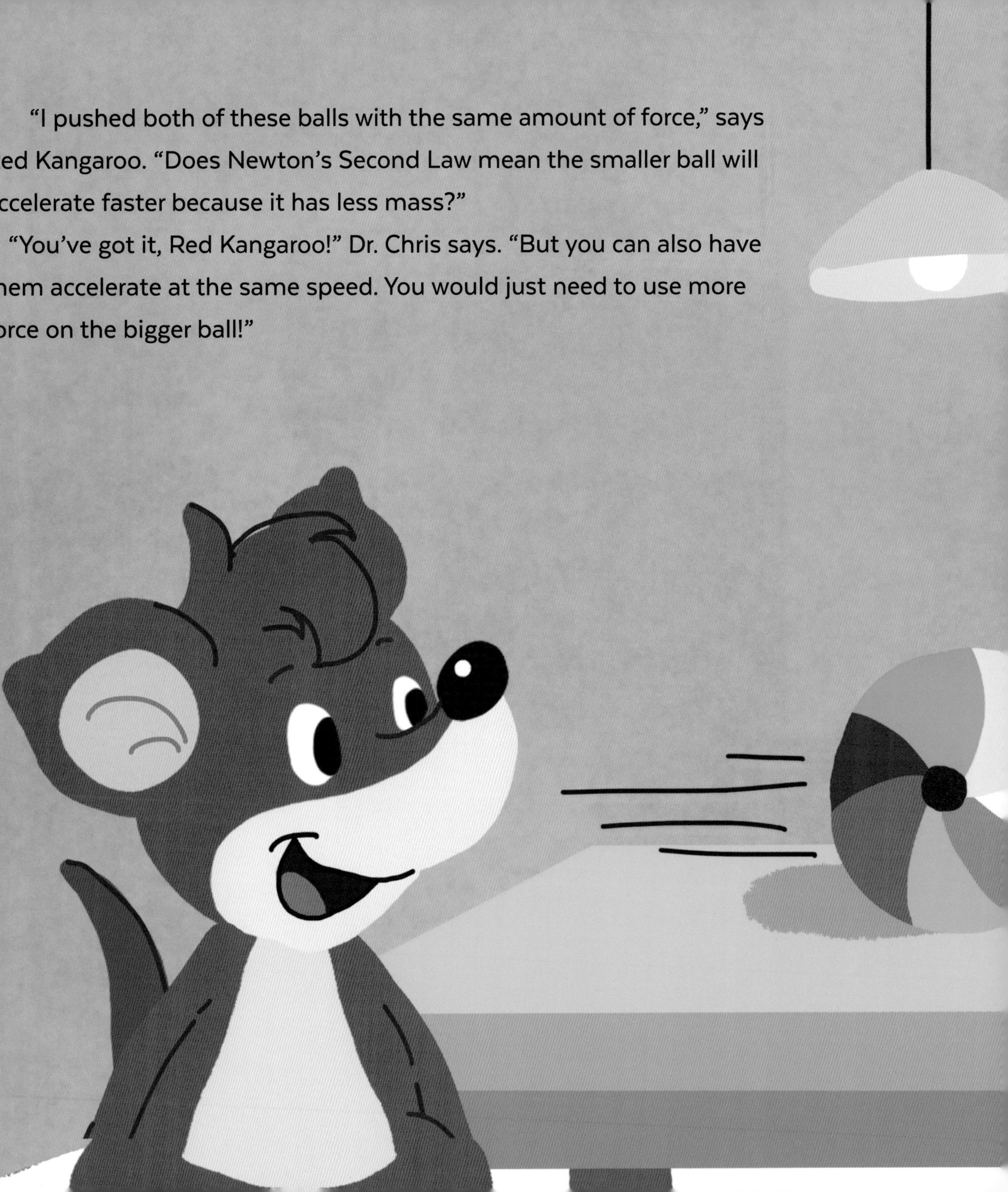

Dr. Chris

3

“We just have one law left! **Newton’s Third Law** says that for every force that acts on an object, there is an equal force acting in the opposite direction.”

"When I push on you, Dr. Chris, I can feel you pushing back on me!" Red Kangaroo says.

"Exactly!" says Dr. Chris. "And you feel the weight of your ball pushing back on your hand when you push it or throw it in the air!"

“Thanks for teaching me the rules of **motion**, Dr. Chris!” says Red Kangaroo. “Now I can think just like Newton and see his three laws of motion whenever I see something moving!”

Glossary

Acceleration
The change in how fast or how slowly an object is moving.

Force
Any push or pull on an object. Forces are measured in Newtons, in honor of the person who made many discoveries involving the different types of forces.

Gravity
A force that pulls two objects toward each other. Earth's gravity is what makes things fall and what keeps you on the ground.

Mass
The measure of how much matter an object is made of.

Motion
The way something moves in nature.

Newton's First Law of Motion
An object will either stay still or continue moving unless another force affects it.

Newton's Second Law of Motion

The greater the mass, the higher the force needed for acceleration.

Newton's Third Law of Motion

For every force on an object, there is an equal force acting in the opposite direction.

Sir Isaac Newton

A famous physicist who created the laws of motion. Some people believe he understood the force of gravity after an apple fell off a tree and hit him in the head. What do you think?

Show What You Know

1. List the three words Red Kangaroo needed to know before she could understand Newton's Laws of Motion.
2. Name the force that makes things fall.
3. Describe what it means for something to accelerate.
4. The opposite of acceleration is deceleration. Based on what you know about acceleration, explain what happens when an object decelerates.
5. True or False: A car would need more force to move than a bicycle. Explain the answer using your knowledge of Newton's Laws.

Answers on the last page.

Test It Out

Which will fall the fastest?

1. Find two balls that are different sizes, like a basketball and a baseball. Make predictions about which one you think has the bigger mass and which one you think would hit the ground first if you dropped them.

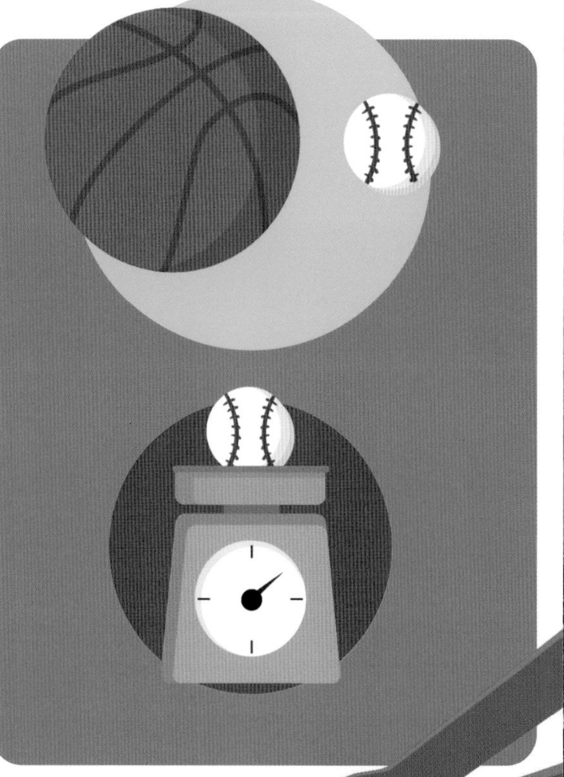

2. A kitchen or bathroom scale can be used to measure the mass of each ball. If you can't find a scale, you can use your hands and arms to decide which is heavier and therefore has more mass.

3. Drop both balls from the same height at the same time (you may need some help from a friend or an adult). If you can't drop both balls at the same time, use a timer to record how much time it takes each ball to hit the ground.

4. Record the results. Were your predictions correct?

Try the experiment again with as many different-sized balls as you can find!

Demonstrate Newton's First Law!

1. Gather a rock, a piece of cardboard, and a cup. Place the cardboard on top of the cup and the rock on the cardboard.
2. Make two predictions: What will happen if you pull the cardboard off the cup slowly? What will happen if you pull the cardboard off the cup quickly?
3. Start by pulling the cardboard quickly. Make sure to pull the cardboard horizontally and not up or down. Record what happened to the rock when you pulled out the cardboard. Did this match your prediction?
4. Now pull the cardboard slowly. Record what happened to the rock now. Was it different from when you pulled it quickly? Did it match your prediction?
5. Explain how Newton's Laws make the rock act like it does. What force or forces are causing the rock to act in this way?

Try the experiment with different materials (paper, plastic, marbles, small toys) to see if the results change.

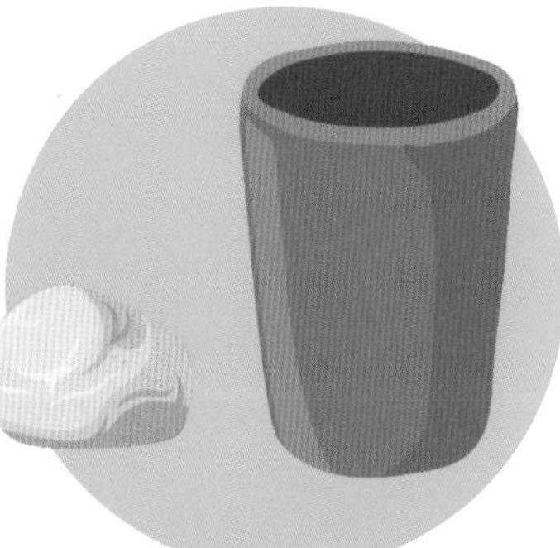

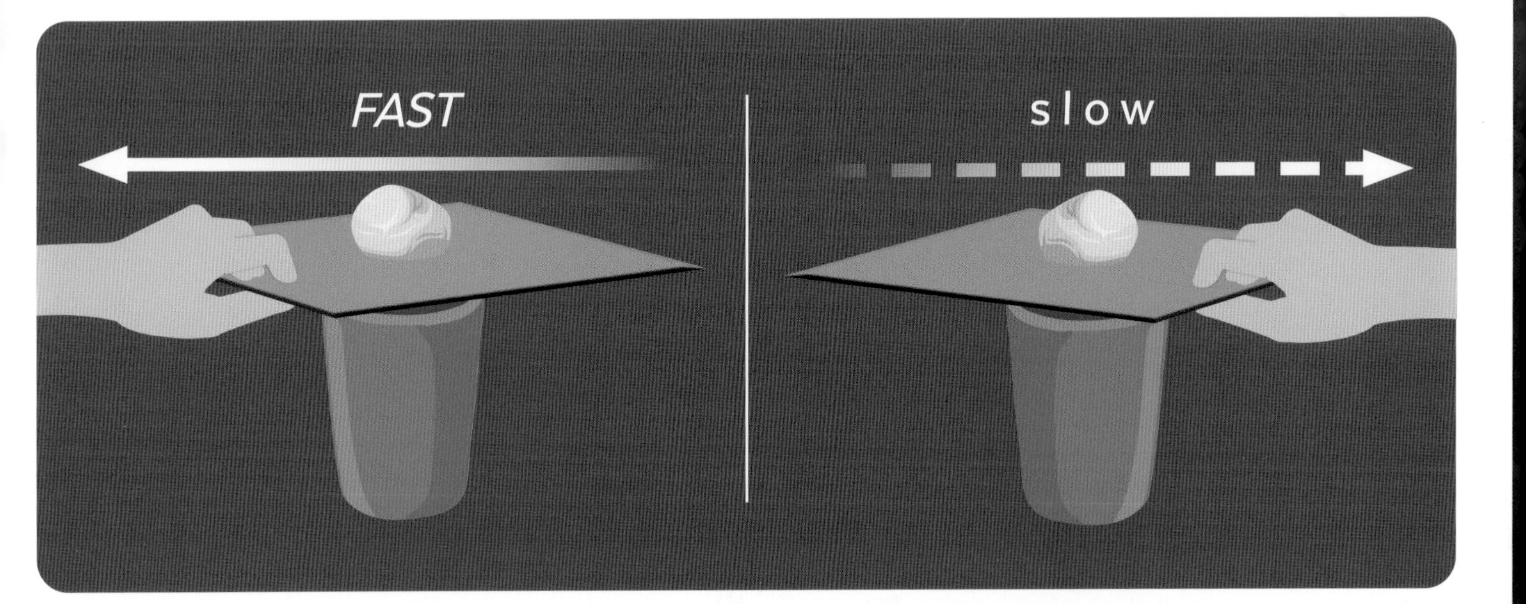

What to expect when you Test It Out

Which will fall the fastest?

The balls should hit the ground at the same time. This is because the bigger ball feels more force from gravity and would need more force to speed up. The smaller ball feels less force from gravity and needs less force to speed up. This makes the balls accelerate at the same speed. Everything falls with the same acceleration unless some other force gets in the way.

Demonstrate Newton's First Law!

The rock will fall into the cup if you pull the cardboard quickly. This happens because of Newton's First Law. The force of gravity pulls the rock down. But the rock does not fall into the cup when you pull the cardboard slowly. This is because of the force of friction keeps the rock moving with the cardboard.

Show What You Know answers

1. Force, mass, and acceleration.
2. Gravity.
3. It moves faster. (Technically: it changes speed or direction.)
4. It slows down. (Technically: it's the negative of acceleration.)
5. True. Newton's Second Law says more mass needs more force to accelerate.